Ozone Layer Depletion

Dr. Hemant Pathak

Copyright © 2018 Hemant Pathak

All rights reserved.

ISBN: 1987691202
ISBN-13: 978-1987691207

DEDICATION

Dedicated to Shri Sainath Maharaj the all omnipotent of world the most merciful

Contents

	Foreword	Vi
	Glossary	9
1	Introduction	18
2	Ultraviolet radiations and their Biological and Abiological Effects	19
3	Importance of ozone Layer	24
4	The Chapman Cycle	25
5	Chemistry of Ozone Depletion	27
6	Causes of Ozone hole over Antarctica	29
7	Ozone Depleting Substances	30
8	Conclusion	32
9	References	34

Foreword

The ozone layer was discovered in 1913 by the French physicists Charles Fabry and Henri Buisson. The ozone layer has the capability to absorb almost 97-99% of the harmful ultraviolet radiations that sun emit and which can produce long term devastating effects on humans beings as well as plants and animals. The earth's stratospheric ozone layer plays a critical role in absorbing ultraviolet radiation emitted by the sun. In the last thirty years, it has been discovered that stratospheric ozone is depleting as a result of anthropogenic pollutants.

Ozone layer depletion is one of the most serious problems faced by our planet earth. It is also one of the prime reasons which are leading to global warming. Ozone Layer depletion describes two related phenomena observed since the late 1970s: a steady decline of about four percent in the total amount of ozone in Earth's stratosphere, and a much larger springtime decrease in stratospheric ozone around Earth's polar regions. The latter phenomenon is referred to as the ozone hole.

Chlorofluorocarbons (CFCs) and other halogenated ozone depleting substances (ODS) are mainly responsible for man-made chemical ozone depletion.

This book describes of international efforts to protect the ozone layer, the greatest success yet achieved in managing human impacts on the global environment. The book provides an account of the ozone-depletion issues from the first attempts to develop international action in the 1970s to the mature functioning of the montreal regime.

This Book represent state of knowledge regarding examines the parallel developments of politics and negotiations, scientific understanding and controversy, technological progress, and industry strategy to draws some conclusions concerning the setting of goals for that shaped the issue's development and its effective management.

Simply explained, Ozone layer depletion is an important book bringing together diverse viewpoints from Environmentalist, state agencies and regulators, for all who wish to save Earth with quality life.

Dr. Hemant Pathak

M.Sc. (Gold medalist), Ph. D.

Assistant Professor of Engineering Chemistry

Indira Gandhi Govt. Engineering college, Sagar, MP, India

Acronyms and Symbol

CFC: Chlorofluorocarbon

Class I Substance: One of several groups of chemicals with an ozone-depletion potential of 0.2 or higher. It include include CFCs, halons, carbon tetrachloride, and methyl chloroform. EPA later added HBFCs and methyl bromide to the list by regulation. Lists of class I substances with their ODPs and CAS numbers are available.

Class II Substance: a chemical with an ozone-depletion potential of less than 0.2. all of the HCFCs are class II substances. Lists of class II substances with their ODPs and CAS numbers are available.

DU: Dobson Unit (a measurement of ozone levels)

GWP: Global Warming Potential

- **NO_2:** Nitrogen dioxide
- **O_3:** Ozone
- **ODS:** Ozone-Depleting Substance
- **ODP:** Ozone Depletion Potential
- **PAN:** Peroxyacetyl nitrate
- **PM:** Particulate matter
- **UV:** Ultraviolet Radiation

Glossary	
Air Pollution	The presence of substances in the atmosphere, particularly those that do not occur naturally. These substances are generally contaminants that substantially alter or degrade the quality of the atmosphere. The term is often used to identify undesirable substances produced by human activity, that is, anthropogenic air pollution. Air pollution usually designates the collection of substances that adversely affects human health, animals, and plants; deteriorates structures; interferes with commerce; or interferes with the enjoyment of life.
Aerosol	small droplet or particle suspended in the atmosphere, typically containing sulfur
Atmosphere	A gaseous envelope gravitationally bound to a celestial body (e.g., a planet, its satellite, or a star).
Carbon Tetrachloride	Carbon tetrachloride was widely used as a raw material in many industrial uses, including the production of CFCs, and as a solvent. Solvent use ended when it was discovered to be carcinogenic. It is also used as a catalyst to deliver chlorine ions to certain processes. Its ozone depletion potential is 1.2
Chlorofluorocarbon (CFC)	CFCs are very stable in the troposphere. They are broken down by strong ultraviolet light in the

	stratosphere and release chlorine atoms that then deplete the ozone layer. CFCs are commonly used as refrigerants, solvents, and foam blowing agents. The most common CFCs are CFC-11, CFC-12, CFC-113, CFC-114, and CFC-115. Their ozone depletion potentials are, respectively, 1, 1, 0.8, 1, and 0.6.
Carbon dioxide	Colorless gas, formula CO_2, molecular weight 44; the fourth most abundant gas in dry air.
Dobson Unit	If 100 DU of ozone were brought to the Earth's surface, it would form a layer 1 millimeter thick. In the tropics, ozone levels are typically between 250 and 300 DU year-round. In temperate regions, seasonal variations can produce large swings in ozone levels. For instance, measurements in Leningrad have recorded ozone levels as high as 475 DU and as low as 300 DU. These variations occur even in the absence of ozone depletion, but they are well understood. Ozone depletion refers to reductions in ozone below normal levels after accounting for seasonal cycles and other natural effects.
Ecosystem	An interactive system that includes the organisms of a natural community association together with their abiotic physical, chemical, and geochemical environment.

Environmental policy	A policy initiative aimed at addressing environmental problems and challenges.
Fog	Water droplets suspended in the atmosphere in the vicinity the earth's surface that affect visibility.
Global warming	increase in the average temperature of the earth's surface.
Global Warming Potential	The GWP is the ratio of the warming caused by a substance to the warming caused by a similar mass of carbon dioxide. Thus, the GWP of CO2 is defined to be 1.0 . CFC-12 has a GWP of 8,500, while CFC-11 has a GWP of 5,000. Various HCFCs and HFCs have GWPs ranging from 93 to 12,100. Water, a substitute in numerous end-uses, has a GWP of 0.
Halon	The halons are used as fire extinguishing agents, both in built-in systems and in handheld portable fire extinguishers. Halon production in the U.S. ended on 12/31/93 because they contribute to ozone depletion. They cause ozone depletion because they contain bromine. Bromine is many times more effective at destroying ozone than chlorine. At the time the current U.S. tax code was adopted, the ozone depletion potentials of halon 1301 and halon 1211 were observed to be 10 and 3, respectively. These values are used for tax calculations. Recent scientific studies, however, indicate that the ODPs are at least

	13 and 4, respectively. Note: technically, all compounds containing carbon and fluorine and/or chlorine are halons, but in the context of the Clean Air Act, "halon" means a fire extinguishing agent as described above.
Hydrocarbons	Strictly speaking, organic molecules consisting of just carbon and hydrogen; often loosely applied also to derivatives of hydrocarbons containing oxygen, halogens, etc. The atmospheric burden of hydrocarbons is provided from both natural and anthropogenic emissions.
Hydrochlorofluorocarbon (HCFC)	The HCFCs are one class of chemicals being used to replace the CFCs. They contain chlorine and thus deplete stratospheric ozone, but to a much lesser extent than CFCs. HCFCs have ozone depletion potentials (ODP)ranging from 0.01 to 0.1. Production of HCFCs with the highest ODPs will be phased out first, followed by other HCFCs.
Montreal Protocol	The Montreal Protocol, signed in 1987, governs the end of production of ozone-depleting substances. Under the Protocol, various research groups continue to assess the ozone layer through a series of reports. In addition, the Multilateral Fund provides resources to developing nations to promote the transition to ozone-safe technologies.

Methyl Bromide (CH3Br)	Methyl Bromide's chemical formula is CH3BR. An effective pesticide, this compound is used to fumigate soil and many agricultural products. Because it contains bromine, it depletes stratospheric ozone and has an ozone depletion potential of 0.6.
Nanometer	The nanometer, or nm, is a common unit used to describe wavelengths of light or other electromagnetic radiation such as UV. For example, green light has wavelengths of about 500-550 nm, while violet light has wavelengths of about 400-450 nm. One billionth is a tiny number. One foot is about one billionth the distance of 48 round-trips between Los Angeles and Washington, DC.
Ozone	A nearly colorless gas, formula O3, molecular weight 48, that appears blue in the condensed phase or at high concentration, with a characteristic odor like that of weak chlorine. It is formed in the reaction between atomic oxygen and molecular oxygen
Oxidants	Substance capable of causing oxidation of, for example, an atmospheric species.
ODS	ODS include CFCs, HCFCs, halons, methyl bromide, carbon tetrachloride, and methyl chloroform. ODS are generally very stable in the troposphere and only degrade under intense

	ultraviolet light in the stratosphere. When they break down, they release chlorine or bromine atoms, which then deplete ozone. A detailed list of class I and class II substances with their ODPs and CAS numbers are available.
Particulates	The term for solid or liquid particles found in the air.
Ozone Depletion Potential	The ODP is the ratio of the impact on ozone of a chemical compared to the impact of a similar mass of CFC-11. Thus, the ODP of CFC-11 is defined to be 1.0. Other CFCs and HCFCs have ODPs that range from 0.01 to 1.0. The halons have ODPs ranging up to 10. Carbon tetrachloride has an ODP of 1.2, and methyl chloroform's ODP is 0.11. HFCs have zero ODP because they do not contain chlorine.
Ozone layer	The ozone layer lies approximately 15-40 kilometers (10-25 miles) above the Earth's surface, in the stratosphere. Depletion of this layer by ODS will lead to higher UVB levels, which in turn will cause increased skin cancers and cataracts and potential damage to some marine organisms, plants, and plastics.
Pollutants (pollution)	Unwanted chemicals or other materials found in the air. Pollutants can harm health, the environment and property. Many air pollutants occur as gases or vapors, but some are very tiny solid particles: dust,

	smoke, or soot.
Point pollution	polluted water from a defined point. It can be collected as industrial or municipal wastewater and treated by what is often called end-of-pipe technology (environmental technology).
Pollution control	The addition of processes, practices, materials, products or energy to waste streams to reduce the risk posed by pollutants and waste before their release to the environment.
Pollution prevention	The use of processes, practices, materials, products, substances or energy that avoid or minimize the creation of pollutants and waste, and reduce the overall risk to human health or the environment
Public health	The health or physical well-being of a whole community.
Secondary pollutants	Pollutants that are formed in the atmosphere as a result of chemical reactions. Secondary pollutants are often photochemical oxidants such as ozone or nitrogen dioxide, or components of acid rain such as sulfuric acid or nitric acid.
Stratosphere	The stratosphere extends from about 10km to about 50km in altitude. Commercial airlines fly in the lower stratosphere. The stratosphere gets warmer at

	higher altitudes. In fact, this warming is caused by ozone absorbing ultraviolet radiation. Warm air remains in the upper stratosphere, and cool air remains lower, so there is much less vertical mixing in this region than in the troposphere.
Threatened species	Species of flora or fauna likely to become endangered within the foreseeable future.
Troposphere	The troposphere extends from the surface up to about 10 km in altitude, although this height varies with latitude. Almost all weather takes place in the troposphere. Mt. Everest, the highest mountain on Earth, is only 8.8 km high. Temperatures decrease with altitude in the troposphere. As warm air rises, it cools, falling back to Earth. This process, known as convection, means there are huge air movements that mix the troposphere very efficiently.
Ultraviolet radiation	Ultraviolet radiation is a portion of the electromagnetic spectrum with wavelengths shorter than visible light. The sun produces UV, which is commonly split into three bands: UVA, UVB, and UVC. UVA is not absorbed by ozone. UVB is mostly absorbed by ozone, although some reaches the Earth. UVC is completely absorbed by ozone.
UVA	UVA is not absorbed by ozone. This band of radiation has wavelengths just shorter than visible

	violet light.
UVB	UVB is a kind of ultraviolet light that is particularly effective at damaging DNA. It is a cause of melanoma and other types of skin cancer. It has also been linked to damage to some materials, crops, and marine organisms. The ozone layer protects the Earth against most UVB coming from the sun. It is always important to protect oneself against UVB, even in the absence of ozone depletion, by wearing hats, sunglasses, and sunscreen.
UVC	UVC is extremely dangerous, but it is completely absorbed by ozone and normal oxygen (O_2)

1. Introduction

Ozone layer depletion is one of the most serious problems faced by our planet earth. It is also one of the prime reasons which are leading to global warming. Ozone Layer depletion describes two phenomena observed since the late 1970s: a steady decline of about 4% in the total amount of ozone in Earth's stratosphere and a much larger springtime decrease in stratospheric ozone around Earth's polar regions.

The latter phenomenon is referred to as the ozone hole. The ozone layer is a belt of the naturally occurring gas ozone. It exists 16 to 32 kilometers above earth, and serves as a shield from the harmful ultraviolet radiation emitted by the sun. Ozone layer absorbs these harmful radiations and thus prevents these rays from entering the earth's atmosphere.

Ozone is a colourless gas found in the upper atmosphere of the Earth are highly reactive molecule that contains three oxygen atoms. It is formed when oxygen molecules absorb ultraviolet photons, and undergo a chemical reaction known as photo dissociation or photolysis. It is constantly being formed and broken down in the high atmosphere, 10 to 50 kilometers above Earth, in the region called the stratosphere. Ozone molecule has ability to block solar radiations of wavelengths less than 290 nanometers from reaching Earth's surface. In this process, it also absorbs ultraviolet radiations that are dangerous for most living beings.

The main cause of ozone depletion and the ozone hole is man-made chemicals, especially man-made halocarbon refrigerants, solvents, propellants, and foam-blowing agents, referred to as ozone depleting substances or ODS.

These compounds are transported into the stratosphere by the winds after being emitted at the surface. Once in the stratosphere, they release halogen atoms through photo-dissociation, which catalyze the breakdown of ozone into oxygen. Both types of ozone depletion were observed to increase as emissions of halocarbons increased.

From the 1970s the depletion of the ozone layer started to capture the attention of the scientists, environmentalists, and the world community at large. These concerns led to the adoption of the Montreal Protocol in 1987, which bans the production of CFCs, halons, and other ozone-depleting chemicals.

Ozone depletion and the ozone hole have generated worldwide concern over increased cancer risks and other negative effects.

There are many other substances that lead to ozone layer depletion such as hydro chlorofluorocarbons (HCFCs) and volatile organic compounds (VOCs). These substances are found in vehicular emissions, by-products of industrial processes, aerosols and refrigerants.

All these ozone depleting substances remain stable in the lower atmospheric region, but as they reach the stratosphere, they get exposed to the ultra violet rays. This leads to their breakdown and releasing of free chlorine atoms which reacts with the ozone gas, thus leading to the depletion of the ozone layer.

2. Ultraviolet radiations and their Biological and Abiological Effects

These are high energy electromagnetic waves emitted by the sun which if enters the earth's atmosphere can lead to various environmental issues including global warming, and also a number of health related issues for all living beings.

These wavelengths cause skin cancer, sunburn, and cataracts, which were projected to increase dramatically as a result of thinning ozone, as well as harming plants and animals. Ultraviolet radiation could destroy the organic matter.

Plants and plankton cannot thrive, both acts as food for land and sea animals, respectively. For humans, excessive exposure to ultraviolet radiation leads to higher risks of cancer and cataracts. UV rays will also affect the plants. It can also directly affect the plant growth by altering the physiological and developmental processes of the plants. In some species of plants, UV radiation can alter the time of flowering, as well as the number of flowers produced by a plant.

Plant growth can be directly affected by UV-B radiation. Despite mechanisms to reduce or repair these effects, physiological and developmental processes of plants are affected. UVB radiation affects the physiological and developmental processes of plants. Despite mechanisms to reduce or repair these effects and an ability to adapt to increased levels of UVB, plant growth can be directly affected by UVB radiation.

UV radiation includes UV-A, the least dangerous form of UV radiation, UV-B, and UV-C, which is the most dangerous. UV-C is unable to reach the Earth's surface due to stratospheric ozone's ability to absorb it. The real threat comes from UV-B, which can enter the Earth's atmosphere, and has adverse effects.

Ozone layer depletion increases the amount of UVB that reaches the Earth's surface. UVB has been linked to the development of cataracts, a clouding of the eye's lens. The main public concern regarding the ozone hole has been the effects of increased surface UV radiation on human health.

Laboratory studies demonstrate that exposure to UV rays from sun can lead to increased risk for developing of several types of skin cancers. Malignant melanoma, basal and squamous cell carcinoma are the most common cancers caused by exposure to UV rays. UV rays are harmful for our eyes too. Our immune system is also highly sensitive to UV rays. Increased exposure to UV rays can lead to weakening of the response of immune system and even impairment of the immune system in extreme cases.

UV rays can lead to acceleration of the aging process of your skin. Consequently looking older than what you actually are. It can also lead to photo allergy that result in outbreak of rashes in fair skinned people. In humans, exposure to UV rays can also lead to difficulty in breathing, chest pain, and throat irritation and can even lead to hampering of lung function.

UV rays affect other life forms too. It adversely affects the different species of amphibians and is one of the prime reasons for the declining numbers of the amphibian species. It affects them in every stage of their life cycle; from hampering the growth and development in the larvae stage, deformities and decreases immunities in some species and to even retinal damage and blindness in some species.

UV rays also have adverse effect on the marine ecosystem. It adversely affects the planktons which plays a vital role in the food chain and oceanic carbon cycle. Affecting phytoplankton will in turn affect the whole ocean ecosystem.

Institute of Zoology in London found that whales off the coast of California have shown a sharp rise in sun damage, and these scientists fear that the thinning ozone layer is to blame. The findings suggest "rising UV levels as a result of ozone depletion are to blame for the observed skin damage, in the same way that human skin cancer rates have been on the increase in recent decades.

Ozone layer depletion leads to decrease in ozone in the stratosphere and increase in ozone present in the lower atmosphere. Presence of ozone in the lower atmosphere is considered as a pollutant and a greenhouse gas. Ozone in the lower atmosphere contributes to global warming and climate change. The depletion of ozone layer has trickle down effects in the form of global warming, which in turn leads to melting of polar ice, which will lead to rising sea levels and climatic changes around the world.

An increase of UV radiation would be expected to affect crops. A number of economically important species of plants, such as rice, depend on cyanobacteria residing on their roots for the retention of nitrogen. Cyanobacteria are sensitive to UV radiation and would be affected by its increase.

Ozone depletion is listed as one of the causes for the declining numbers of amphibian species. Ozone depletion affects many species at every stage of their life cycle.

- Hampers growth and development in larvae
- Changes behavior and habits
- Causes deformities in some species
- Decreases immunity. Some species have become more vulnerable to diseases and death
- Retinal damage and blindness in some species

Phytoplankton form the foundation of aquatic food webs. Phytoplankton productivity is limited to the euphotic zone, the upper layer of the water column in which there is sufficient sunlight to support net productivity. Plankton are threatened by increased UV radiation. Marine phytoplankton play a fundamental role in both the food chain as well as the oceanic carbon cycle. It plays an important role in converting atmospheric carbon dioxide into oxygen. Ultraviolet rays can influence the survival rates of these microscopic organisms, by affecting their orientation and mobility. This eventually disturbs and affects the entire ecosystem.

Exposure to solar UVB radiation has been shown to affect both orientation and motility in phytoplankton, resulting in reduced survival rates for these organisms. Scientists have demonstrated a direct reduction in phytoplankton production due to ozone depletion-related increases in UVB.

UVB radiation has been found to cause damage to early developmental stages of fish, shrimp, crab, amphibians, and other marine animals. The most severe effects are decreased reproductive capacity and impaired larval development. Small increases in UVB exposure could result in population reductions for small marine organisms with implications for the whole marine food chain.

Increases in UVB radiation could affect terrestrial and aquatic biogeochemical cycles, thus altering both sources and sinks of greenhouse and chemically important trace gases. These potential changes would contribute to biosphere-atmosphere feedbacks that mitigate or amplify the atmospheric concentrations of these gases.

Synthetic polymers, naturally occurring biopolymers, as well as some other materials of commercial interest are adversely affected by UVB radiation.

3. Importance of ozone Layer

Ozone layer depletion first captured the attention of the whole world in the 1970, and a lot of research has been done to find its possible effects and causes. Scientists have anticipated disruption of susceptible terrestrial and aquatic ecosystems due to depletion of ozone layer. UV radiation could destroy life on Earth. Absorption of UV radiations warms the stratosphere but it is important for life to flourish on planet Earth.

Low temperatures, increase in the level of chlorine and bromine gases in the upper stratosphere are some of the reasons that leads to ozone layer depletion. But the one and the most important reason for ozone layer depletion is the production and emission of chlorofluorocarbons (CFCs). Chlorofluorocarbons, which can travel to the stratosphere without being destroyed in the troposphere due to their low reactivity

It is calculated that every 1 percent decrease in ozone layer results in a 1-5 percent increase in the occurrence of skin cancer. Other worst effects of the reduction of protective ozone layer include incidence of cataracts, sunburns and weak immune system.

During 1985 to 1988 studied on over the south pole continually noticed significantly reduced concentrations of ozone directly over the continent of Antarctica. an enormous hole in the ozone layer had indeed developed over Antarctica. NASA satellite data later showed that the hole had begun to develop in the mid 1970's.

4. The Chapman Cycle

The stratosphere is in a continues cycle with oxygen molecules and their interaction with UV rays. This process is considered a cycle because of its constant conversion between different molecules of oxygen. In this cycle, a single molecule of oxygen breaks down into two oxygen atoms. The free oxygen atom (O), then combines with an oxygen molecule (O_2), and forms a molecule of ozone (O_3). The ozone molecules, in turn absorb ultraviolet rays between 310 to 200 nm (nanometers) wavelength, and thereby prevent these harmful radiations from entering the Earth's atmosphere. The process of absorption of harmful radiation occurs when ozone molecules split up into a molecule of oxygen, and an oxygen atom. The oxygen atom, recombines with the oxygen molecule (O_2) to regenerate an ozone (O_3) molecule. Thus, the total amount of ozone is maintained by this continuous process of destruction, and regeneration.

Three forms oxygen are involved in the ozone-oxygen cycle: oxygen atoms, oxygen gas and ozone gas(O_3). Ozone is formed in the stratosphere when oxygen molecules photodissociate after intaking ultraviolet photons. This converts a single O_2 into two atomic oxygen radicals. The atomic oxygen radicals then combine with separate O_2 molecules to create two O_3 molecules. These ozone molecules absorb ultraviolet (UV) light, following which ozone splits into a molecule of O_2 and an oxygen atom. The oxygen atom then joins up with an oxygen molecule to regenerate ozone. This is a continuing process that terminates when an oxygen atom recombines with an ozone molecule to make two O_2 molecules.

The ozone layer is created when ultraviolet rays react with oxygen molecules (O_2) to create ozone (O_3) and atomic oxygen (O). This process is called the Chapman cycle.

Step 1: An oxygen molecules is photolyzed by solar radiation, creating two oxygen radicals:

$$h\nu + O_2 \rightarrow 2O.$$

Step 2: Oxygen radicals then react with molecular oxygen to produce ozone:

$$O_2 + O. \rightarrow O_3$$

Step 3: Ozone then reacts with an additional oxygen radical to form molecular oxygen:

$$O_3 + O. \rightarrow 2O_2$$

Step 4: Ozone can also be recycled into molecular oxygen by reacting with a photon:

$$O_3 + h\nu \rightarrow O_2 + O.$$

Ozone is constantly being created and destroyed by the Chapman cycle and that these reactions are natural processes, which have been taking place for millions of years.

Thickness the ozone layer at any particular time can vary greatly. O_2 is constantly being introduced into the atmosphere through photosynthesis, so the ozone layer has the capability of regenerating itself.

5. Chemistry of Ozone Depletion

Ozone layer can be destroyed by a number of free radical catalysts; the important are the hydroxyl radical (OH·), nitric oxide radical (NO·), chlorine radical (Cl·) and bromine radical (Br·). The dot indicate that each species has an unpaired electron and is thus extremely reactive. All of these have both natural and man-made sources, most of the OH· and NO· in the stratosphere is naturally occurring, but anthropogenic activity has drastically increased the levels of chlorine and bromine.

CFC molecules are made up of chlorine, fluorine and carbon atoms and are extremely stable. Chlorofluorocarbons (CFCs), chemicals found mainly in spray aerosols heavily used by industrialized nations for much of the past 60 years, are the primary culprits in ozone layer breakdown.

Cl and Br atoms are released from the parent compounds by the action of ultraviolet light, e.g.

$$CFCl_3 + \text{electromagnetic radiation} \rightarrow Cl\cdot + \cdot CFCl_2$$

Ozone is a highly reactive molecule that easily reduces to the more stable oxygen form with the assistance of a catalyst. Cl and Br atoms destroy ozone molecules through a variety of catalytic cycles.

Chlorine atom reacts with an ozone molecule (O_3), taking an oxygen atom to form chlorine monoxide (ClO) and leaving an oxygen molecule (O_2). The ClO can react with a second molecule of ozone, releasing the chlorine atom and yielding two molecules of oxygen. The chemical shorthand for these gas-phase reactions is:

- $Cl\cdot + O_3 \rightarrow ClO + O_2$

A chlorine atom removes an oxygen atom from an ozone molecule to make a ClO molecule.

- $ClO + O_3 \rightarrow Cl\cdot + 2 O_2$

This ClO can also remove an oxygen atom from another ozone molecule; the chlorine is free to repeat this two-step cycle.

$$O_3 + O. \rightarrow 2O_2$$

The overall effect is a decrease in the amount of ozone, though the rate of these processes can be decreased by the effects of null cycles. This extreme stability allows CFC's to slowly make their way into the stratosphere. Most of the molecules decompose before they can cross into the stratosphere from the troposphere. This reason the atmosphere allows them to reach great altitudes where photons are more energetic. When the CFC's come into contact with these high energy photons, their individual components are freed from the whole.

Overall reaction expressed, Chlorine is able to destroy most of the ozone because it acts as a catalyst. Chlorine initiates the breakdown of ozone and combines with a freed oxygen to create two oxygen molecules. After each reaction, chlorine begins the destructive cycle again with another ozone molecule. One chlorine atom can thereby destroy thousands of ozone molecules. Because ozone molecules are being broken down they are unable to absorb any ultraviolet light so more intense UV radiation is at the earths surface. A single chlorine atom is able to react with an average of 100,000 ozone molecules before it is removed from the catalytic cycle.

6. Causes of Ozone hole over Antarctica

The ozone hole over Antarctica is formed by unique atmospheric conditions over the continent that combine to create an ideal environment for ozone destruction. The ozone hole is usually measured by reduction in the total column ozone above a point on the Earth's surface. This is normally expressed in Dobson units (DU). The most prominent decrease in ozone has been in the lower stratosphere.

1. Antarctica is surrounded by water, winds over the continent blow in a unique clockwise direction creating a so called "polar vortex" that effectively contains a single static air mass over the continent. Consequently air over Antarctica does not mix with air in the rest of the earth's atmosphere.

2. Antarctica has the coldest winter temperatures on earth, often reaching -110 F. These chilling temperatures result in the formation of polar stratospheric clouds (PSC's) which are a conglomeration of frozen H_2O and HNO_3. Due to their extremely cold temperatures, polar stratospheric clouds form an electrostatic attraction with CFC molecules as well as other halogenated compounds

3. During spring season in Antarctica, the PSC's melt in the stratosphere and release all of the halogenated compounds that were previously absorbed to the cloud. In the antarctic summer, high energy photons are able to photolyze the halogenated compounds, freeing halogen radicals that then catalytically destroy O_3.

4. Antarctica is constantly surrounded by a polar vortex, radical halogens are not able to be diluted over the entire globe. The ozone hole develops as result of this process.

5. Research suggests that the strength of the polar vortex from any given year is directly correlated to the size of the ozone hole. In years with a strong polar vortex, the ozone hole is seen to expand in diameter, whereas in years with a weaker polar vortex, the ozone hole is noted to shrink.

Some people thought that the ozone hole should be above the sources of CFCs. However, CFCs are well mixed globally in the troposphere and stratosphere. The reason for occurrence of the ozone hole above Antarctica is not because there are more CFCs concentrated but because the low temperatures help form polar stratospheric clouds.

7. **Ozone Depleting Substances**

- **Chlorofluorocarbons (CFCs)**

CFCs were invented by Thomas Midgley, Jr. in the 1920s. They were used in air conditioning and cooling units, as aerosol spray propellants prior to the 1970s, and in the cleaning processes of delicate electronic equipment. It was utilized as a coolant in home appliances like freezers, refrigerators and air conditioners in both buildings and cars that were manufactured prior to 1995. This substance is usually contained in dry cleaning agents, hospital sterilants, and industrial solvents. The substance is also utilized in foam products like mattresses and cushions and home insulation.

They also occur as by-products of some chemical processes. No significant natural sources have ever been identified for these compounds, its presence in the atmosphere is due almost entirely to human manufacture.

when this chemicals reach the stratosphere, dissociated by ultraviolet light to release chlorine atoms. The chlorine atoms act as a catalyst, and each can break down tens of thousands of ozone molecules before being removed from the stratosphere. Given the longevity of CFC molecules, recovery times are measured in decades. It is calculated that a CFC molecule takes an average of about five to seven years to go from the ground level up to the upper atmosphere, and it can stay there for about a century, destroying up to one hundred thousand ozone molecules during that time.

It's billed as the most extensively utilized ozone-depleting substance because it attributes to more than 80% of overall ozone depletion

- **Hydrofluorocarbons (HCFCs)**

Hydrofluorocarbons have over the years served in place of Chlorofluorocarbons. They are not as harmful as CFCs to ozone layer.

- **Halons**

It's especially used in selected fire extinguishers in scenarios where the equipment or material could be devastated by water or extinguisher chemicals.

- **Carbon Tetrachloride**

Widely used in selected fire extinguishers and solvents.

- **Methyl Chloroform**

Commonly utilized in industries for cold cleaning, vapor degreasing, chemical processing, adhesives and some aerosols.

8. Conclusion

The Antarctic ozone hole is an area of the Antarctic stratosphere in which the recent ozone levels have dropped to as low as 33 percent of their pre-1975 values. The ozone hole occurs during the Antarctic spring, from September to early December, as strong westerly winds start to circulate around the continent and create an atmospheric container. Within this polar vortex, over 50 percent of the lower stratospheric ozone is destroyed during the Antarctic spring.

A gradual trend toward healing was reported in 2016. In 2017, NASA announced that the ozone hole was the weakest since 1988 because of warm stratospheric conditions. It is expected to recover around 2070.

The amount lost is more variable year-to-year in the Arctic than in the Antarctic. The greatest Arctic declines are in the winter and spring, reaching up to 30 percent when the stratosphere is coldest.

There are some ideal solutions to stop Ozone Depletion are as follows-

Pesticides are great chemicals to rid your farm of pests and weeds, but they contribute to ozone layer depletion. Solution is to get rid of pests and weeds is to apply natural methods. Just weed your farm manually and use alternative eco-friendly chemicals to alleviate pests.

The easiest technique to minimize ozone depletion is to limit the number of vehicles on the road. These vehicles emit a lot of greenhouse gases that eventually form smog, a catalyst in the depletion of ozone layer.

Most household cleaning products are loaded with harsh chemicals contributing to degradation of the ozone layer. Use natural and environmentally friendly cleaning products to arrest this situation. The Montreal Protocol formed in 1989 helped a lot in the limitation of Chlorofluorocarbons (CFCs). Nitrous oxide is still in use today. Governments must take action now and outlaw nitrous oxide use to reduce the rate of ozone depletion.

In 1994, the United Nations General Assembly voted to designate September 16 as the International Day for the Preservation of the Ozone Layer, or *World Ozone Day*, to commemorate the signing of the Montreal Protocol on that date in 1987.

9. References

1. Dessler, Andrew. The Chemistry and Physics of Stratospheric Ozone. San Diego, Ca: Academic Press, 2000

2. Hoffman, Matthew J. Ozone Depletion and Climate Change. Albany, NY: State University of New York Press, 2005

3. Parson, Edward A. Protecting the Ozone Layer: Science and Strategy. New York: Oxford University Press, 2003.

4. Petrucci, Ralph H., William S. Harwood, and Geoff E. Herring. General Chemistry : Principles and Modern Applications. 9th ed. Upper Saddle River: Prentice Hall, 2006.

5. Varotsos, Costas, Kirill Ya. Kondratyev. Atmospheric Ozone Variability: Implications for Climate Change, Human Health and Ecosystems. Chichester, UK: Praxis Publishing Ltd, 2000

6. Godish, Thad. Air Quality. 4th ed. Florida: CRC Press LLC, 2004.

7. United States Clean Air Act: as of June 3rd, 2010.

ABOUT THE AUTHOR

Dr. Hemant Pathak held positions as Assistant Professor in the department of chemistry, Govt. Indira Gandhi Engineering College, Sagar, MP, India. He had extensive experience in teaching, research and administrative management.

Dr. Pathak received his Ph.D. degree in chemistry from Dr. Hari Singh Gour Central University, Sagar, India and M.Sc. Gold medalist from Jiwaji University, Gwalior. He has published 29 books and more than 50 research papers in reputed International and National journals and received several awards. He is a member of editorial boards and reviewer boards of several international journals and societies. His area of specialization includes Engineering Chemistry, Energy audits and Environmental Pollution management.

www.ingramcontent.com/pod-product-compliance
Lightning Source LLC
Chambersburg PA
CBHW062236220526
45471CB00009B/3503